BULLHEAD SHARKS

by Julie Murray

Cody Koala
An Imprint of Pop!
popbooksonline.com

Hello! My name is Cody Koala

This book is filled with videos, puzzles, games, and more! Scan the QR codes* while you read, or visit the website below to make this book pop.

popbooksonline.com/bullhead

*Scanning QR codes requires a web-enabled smart device with a QR code reader app and a camera.

abdobooks.com

Published by Pop!, a division of ABDO, PO Box 398166, Minneapolis, Minnesota 55439. Copyright ©2024 by Abdo Consulting Group, Inc. International copyrights reserved in all countries. No part of this book may be reproduced in any form without written permission from the publisher. Cody Koala™ is a trademark and logo of Pop!.

Printed in the United States of America, North Mankato, Minnesota.
052023
082023
THIS BOOK CONTAINS RECYCLED MATERIALS

Cover Photo: Blue Planet Archive
Interior Photos: Shutterstock Images; Getty Images; Blue Planet Archive
Editor: Elizabeth Andrews; Grace Hansen
Series Designer: Victoria Bates

Library of Congress Control Number: 2022950509

Publisher's Cataloging-in-Publication Data
Names: Murray, Julie, author.
Title: Bullhead sharks / by Julie Murray
Description: Minneapolis, Minnesota : Pop!, 2024 | Series: Sharks | Includes online resources and index
Identifiers: ISBN 9781098244224 (lib. bdg.) | ISBN 9781098244927 (ebook)
Subjects: LCSH: Bullhead sharks--Juvenile literature. | Sharks--Juvenile literature. | Sharks--Behavior--Juvenile literature. | Marine fishes--Behavior--Juvenile literature.
Classification: DDC 598.47--dc23

Table of Contents

Chapter 1
Bottom Dwellers 4

Chapter 2
Different Kinds 8

Chapter 3
Nocturnal Animals 14

Chapter 4
Life of a Bullhead Shark . . .18

Making Connections 22
Glossary .23
Index 24
Online Resources 24

Chapter 1

Bottom Dwellers

Bullhead sharks live in **tropical** and **subtropical** waters around the world. They spend most of their time in shallow water.

Watch a video here!

Bullhead sharks are **bottom-dwelling** fish. They are slow and **clumsy** swimmers. They use their fins to push themselves along the ocean floor.

> A bullhead shark will often stay in one area for its entire life.

Chapter 2

Different Kinds

There are several different **species** of bullhead sharks. Most are brown, gray, or black in color. The zebra bullhead shark has dark bands around its body.

Learn more here!

Most bullhead shark species grow to be about 3.3 feet (1m) long. The Mexican horn shark is the largest species of bullhead shark. It can grow up to 5.6 feet (1.7m) long.

> The Japanese bullhead shark can be found in the Pacific Ocean near Japan, North Korea, South Korea, and China. It can grow up to 3.9 feet (1.2m) long.

Bullhead sharks have a **broad** head. They have a heavy brow bone over each eye. This gives them a bull-like

appearance. The sharks also have a short nose and a small mouth.

Chapter 3

Nocturnal Animals

Bullhead sharks are mainly active at night. During the day, they stay hidden in rock **crevices**. They also hide in kelp patches.

Explore links here!

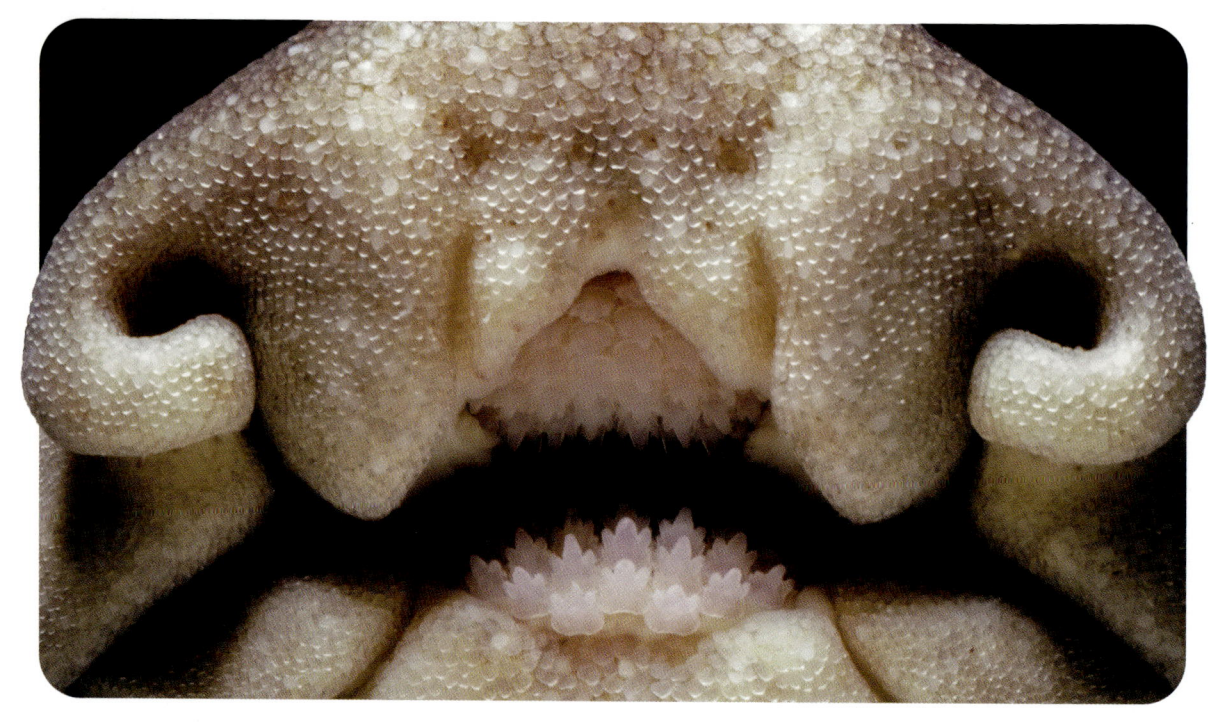

Bullhead sharks have flat back teeth. These are good for cracking the hard shells of shrimp and crabs.

The sharks also eat small fish and sea urchins.

Crested bullhead sharks mainly eat sea urchins. This diet stains their teeth purple!

Chapter 4

Life of a Bullhead Shark

Bullhead sharks lay eggs. The egg cases are spiral shaped. The female lays two eggs at a time. She shoves the eggs into **crevices** to keep them safe from hungry animals.

Complete an activity here!

19

The eggs hatch five to 12 months later. Baby bullhead sharks are 5.5 to 9.5 inches (14 to 24cm) long at birth. Bullhead sharks can live ten to 25 years.

Making Connections

Text-to-Self

Imagine that you are swimming in the ocean and see a bullhead shark. Would you be scared? Why or why not?

Text-to-Text

Have you read a book about a different kind of shark? How is that shark like a bullhead shark? How is it different?

Text-to-World

Bullhead sharks are nocturnal animals. Can you think of another animal that is active at night?

Glossary

bottom-dwelling – living and feeding on or near the bed of a sea, lake, or other body of water.

broad – wide; large.

clumsy – without physical grace or control; awkward.

crevice – a small opening, especially in a rock or a wall.

species – a group of living things that look alike and can have young with one another.

subtropical – relating to an area bordering a tropical area.

tropical – relating to an area with an average temperature above 77 degrees Fahrenheit (25°C) where no freezing occurs.

Index

baby bullhead sharks, 21

body, 11

coloring, 8

eggs, 18, 21

fins, 7, 12

food, 16–17

gills, 12

habitat, 4, 7, 14

head, 12–13

lifespan, 21

size, 11, 21

tail, 12

teeth, 16

Online Resources

popbooksonline.com

Thanks for reading this **Cody Koala** book!

This book is filled with videos, puzzles, games, and more! Scan the QR codes* while you read, or visit the website below to make this book pop.

popbooksonline.com/bullhead

*Scanning QR codes requires a web-enabled smart device with a QR code reader app and a camera.